Lottery Pro Player

Steps To Winning Continuum

Eze Ugbor

I want to dedicate this book to my God for giving me wisdom & strength to make this book possible. I also want to thank my family and friends for being there with me patiently through this process. Finally, I would like to thank my mother for giving me guidance and teaching me how to stay focused even when things seem very difficult.

I thank you all!

Allied Publishing, llc. All Rights Reserved. No part of this material shall be reproduced or transmitted in any form by any means, including photocopying, recording, information retrieval, and by any other means without written permission from the publisher. The numbers in this guide have been carefully worked out with our readers in mind; however, these are recommendations only and does not necessarily guarantee winnings.

All other trademarks or registered trademarks are the property of their respective owners.

Table of Contents

- Chapter 1: Winning Fingertips
- Chapter 2: Enjoy the Ride
- Chapter 3: Examination of Lottery Odds
- Chapter 4: The Cash is in Your Court
- Chapter 5: The Pro Wager
- Chapter 6: Lottery Wagering Know How
- Chapter 7: Mastery Of The Trend Is Winning
- Chapter 8: The Pro T Set System
- Chapter 9: The Pro Source Of Major Money
- Chapter 10: Pick 4 Trend
- Chapter 11: Monthly Pick 4 Templates
- Chapter 12: Cornucopia Methods

Chapter 1

Winning Fingertips

One of the reason people become professionals in any discipline is to make good living from something they excel at. You can become a professional in any field of your choice.

It takes dedication, focus and a lot more to become a pro.

The lottery game is the only game where you have the good fortune of playing and winning like a pro. You have the real fortune of becoming a pro with the right tools.

The job of sleepless nights and working on the numbers is already done for you.

The only thing you need to do is follow the instructions and keep on winning.

This book will deal with pick 4 and pick 3 with major emphasis on pick 4. It will give you examples of the places the numbers played and how to apply them in any state and country. It sure will be a thing of joy to win the lottery even when you are on vacation.

You have to question things, including the lottery odd calculations. You have to identify trends to win consistently. People will not often choose to become professionals if there are no opportunities. The lottery game is no exception.

This book will show you how to win based on your local state lottery. It will show you how to win based on other state lottery results. This is the only book in the market with that level of information.

You do not need more than two-day results to win pick 3 and pick 4 games.

In the mind of the professional player, there are things that must be done. One of those is recognizing the team spirit. The team execution is part of what makes the player. The best players will often stand out from the crowd.

The good player is one that constantly practice, the one that thinks about improvement. The practice includes waking up when others are sleeping and getting the play done.

You are about to enter the world of pro wagering. You are about to excel more than others. You are about to find out that lottery is not just a game but business.

Your beginning of understanding that concept is recognizing that you wager with actual money. That money is cost. You should not continue doing business that costs you money and no return to show for it. This book will be your friend. This book will make you more money than you ever thought possible.

It is high time you consider giving more to your favorite charity.

If you can win only $120000 or $240000 in a year the effort is well worth it. This is not pie in the sky. **You are going to do some work.** Your primary job will be to follow the instructions in this book. I don't believe it is too much to ask for considering the potential winnings.

You will not find the methods in this book in any other lottery books out there.

Chapter 2

Enjoy the Ride

There are several methods to win the lottery games in this book.

You will have trends to win any state or country of your choice. You will have numbers to win any state based on the results from other states. You have ample states to choose from based on trends. There are relationships between numbers being played at any given time among several states. That relationship is a major key to consistent winnings.

Question: Is it possible to win pick 4 everyday?

Answer: It Is Unequivocal Yes.

The answer to winning every pick 4 game.

All pick 4 numbers must travel at certain trend. If and when the trend changes, the calculations must be redone to capture the new trend. You can win pick 4 games where others may not see based on the Cornucopia Methods and the existing trend.

This is possible because the prior results serve as the mirror to the next winners thereby creating relentless opportunities. The parlay method on pick 4 games make the opportunities endless.

I will discuss Cornucopia Methods in later chapters. You will be forever glad with Cornucopia Methods.

There are numbers that trigger others. You will have enough to recognize those and go after the big money.

This book will show you how and when to play parlay. It is a tool that enables you to play with the house money. It is a must for every pro lottery player.

This book will show you how to apply parlay in not just pick 3 but pick 4 as well. The pick 4 is where the real money is and that is where you will enjoy this book even more.

The games here will give you actual examples from different states. I bet you to cross check the examples through any state and country of your choice. Feel free to check the examples all the way to the very first day lottery games became available.

There is no other lottery book that would offer you that invitation.

The pick 4 pays upwards of 5000 percent on your money. If you can win just one pick 4 straight every two weeks that will come out to about $130000. That of course is based on wagering only $1.

You might be wondering if the house will be willing to pay out that much money to you. You should not be concerned about that. The lottery money is as big as the next biggest business out there. There are enough people that do not win. In order to be a pro you have to separate yourself from the crowd.

You have to elevate yourself to the point of asking about the validity of lottery odd calculations. You will master how to wager with patience. You will master execution with precision.

You have to be able to break the odds and do so consistently.

Chapter 3

Examination of Lottery Odds

The popular axiom is that the odd of winning pick 3 is one in one thousand and that of pick 4 is one in ten thousand.

The pick 3 games come in straight, three and six way positions. For instance, 999 can only come in a straight position. 995 is a double game and can only come in three way position as 995, 959 and 599. The six way position is single numbers like 627 that comes in 627, 276, 762, 267, 672 and 726.

The pick 4, on the other hand, comes in four, six, twelve and twenty four way positions. Any of the positions could be won straight. The straight winnings are where the real money is.

In the world of lottery games numbers 1, 2 and 3 does not mean that the next number is going to be 4. In fact, it is very far from being 4. You will question the odd calculations if you understand that. You will win money consistently if you understand that.

Let us take a close look with actual pick 3 games. The word house in this book stands for lottery boards and others responsible for administering the lottery games. If the house plays numbers 123 followed by 234 the average person out there will rush to the local lottery retail outlet to play 345. The reality is that the next number to follow is not 345.

The next number based on that trend is 930. Feel free to check any past lottery results of your choice. This, however, does not mean that 345 is not going to drop. It will play at the right time. The real question then becomes, when is the right time?

The answer is precisely in trend calculations, map of numbers and other formula that will be discussed in later chapters.

If you are playing just pick 3 numbers based on the odd calculation of 1:1000 it will cost you one thousand dollars to win $500 assuming that you don't make any mistake. The alternative is playing over a period of three years to exhaust the one thousand dollars.

The later approach puts you further away from winning anything at all.

The real option available is to either know how to calculate the numbers and come up with the winning 930 in the above instance or have this book and start winning every day.

Do not forget to follow the instructions in the book.

The basket ball players, boxers, musicians and every other field out there do practice. You are not going to win like a pro if you don't put the simple effort of picking the winners based on the instructions in the book.

Let us take a look at one pick 3 number in six way position.

267   672   726   627   276   762

If the pick 3 number 624 played followed by 267 the next pick 3 number based on that trend will be 769.

The same group in this instance could be written as,

624

267

769

The above pick 3 numbers could play in any order. It could start from the top, middle or the last one. The larger point is knowing that the three games move in tandem. There is a key to working out those numbers. You need those keys to win the numbers consistently. The key is what makes you a pro. The key is what separates you from the crowd.

Take a careful look at the above pick 3 numbers.

If, on the other hand, the pick 3 number 862 dropped in place of 624 the entire equation is going to change. This same thing applies to pick 4 games. This book is going to put major emphasis on the pick 4 games. The idea here is for you to consistently win big money. You cannot be a pro without winning major money incessantly.

In the first instance, the trend started with 624. If the 862 played this time followed by 726 the next game to drop will be 171.

The second instance could be written as,

862

726

171

As you can see the equation changed completely because it started with a different pick 3 number.

The above trends could have been playing in two different states. What you need to know is how to use the results of one state to win in your home state. There is no shortage of states to watch.

There are enough trends for you to follow and enable you win serious money.

In another rotation if the 862 played as 287 followed by 726 the third number will completely change although 726 remain the same in the second instance. The third pick 3 number based on that trend will be 995.

The third instance could be written as,

287

726

995

In each of the above instances, the pick 3 numbers are different. This means that the six way position is bound to produce different results. You will follow the trend diligently to get all the money. Please do not leave any money on the table. Professionals do not do that.

As at the time of writing this page, June 6, 2011, let us look at the above groups from one state Texas.

We will explore how to win any state based on results of other states. That is where you will have the chance to make more money than you can imagine.

There are some questions we want to get answers, to since we are examining odd calculations.

Are there relationships on the above instances in the case of Texas lottery?

If there are, did the numbers play in a relative period of time to be profitable?

Did the ones that played give an enough clue to other future opportunities?

If you are privy to the above three instances, would you have made money?

From the eye of Texas lottery Results

The above three instances lined side by side will look like this,

624	862	287
267	726	726
769	171	995

The winnings among the above pick 3 trends will encompass the Texas lottery pick 3 results from 1/1/2011 through 6/6/2011.

Texas played 682 on 1/7/2011 and 711 on 1/18/2011 followed by 278 on 1/22/201 and 627 on 1/25/2011. One month later 679 dropped on 2/28/2011. You will notice that most of the numbers played within a span of one month.

The trend resumed on 3/28/2011 with 276 followed by **624** on 4/9/2011 and 264 on 4/25/2011. The **171** dropped on

5/12/2011 followed by 426 on 5/19/2011 and 627 on 5/26/2011 that repeated as 276 on 5/31/2011.

The underlined pick 3 numbers played straight. The ones that did not play straight so far produced their own respective straight winnings. The number that is yet to play is 995.

There are trends to follow in each case. There are things to look for. The 995 moves with 711 although they are in two different groups. They follow each other and are bound to exhibit the same behavior. If one of the two play five times, the other one is likely going to follow. You can check past results.

The Texas pick 3 lottery results so far gave you twelve numbers won out of three simple trends in less than six months. If we dissect the remaining three you will find out that there is a whole lot more winnings.

We will now begin the fun part and the path to becoming a pro and winning money in the game of lottery.

In later chapters, you are going to get trends that cover extended period of time to enable you win a lot more money. You are going to have the tools to capture the next winners if and when the trend changes. You will get to the pick 4 games. You will begin to have the tools to make money. You will begin to make bank.

There is no other lottery book that can give you the template and know how.

Chapter 4

The Cash is in Your Court

The first thing you should remember is to follow the instructions in this book.

The method here has been tested, fine tuned over thirty years and ready to give you steady stream of cash.

I started working on lottery, since I became a teenager. I embarked on it because I saw too many people lose their hard earned money. The ones that lose are not particularly rich. It bothered me to the point that I decided to work on the numbers and find out every possible angle to winning the lottery.

I started with the English Fixed Odds and eventually embraced every other lottery game.

The major thing is to look at lottery games as business. It is the difference between and the crowd. You are here to wager and win as a pro. You must think through the expense and returns. There is absolutely no need to play the lottery if you do not win consistently.

There are questions you must ask before playing any particular game.

Is there any relationship between the one you plan on wagering and the prior one?

What is the potential risk reward ratio?

Is there the possibility of recouping the cost at a minimum through parlay method?

Is the trend in place to warrant wagering on the game?

Are there precedents to follow on the game?

Is the trend in this book?

If the answers to the above questions and more are not clear, do not wager on that game.

Chapter 5

The Pro Wager

Figure 1

Let us look at some trends and what happened recently. Please remember that you should always remember to be patient. You cannot chase the winning numbers around. The winners will come to you through observation.

A	B	C
122	229	212
555	557	555
393	932	937
623	235	267

471	**715**	**743**
829	**296**	**283**
442	**420**	**448**
962	**627**	**964**
527	**277**	**257**

Let us take a look at Figure 1A through recent Maryland lottery between 6/4/2011 and 6/6/2011. Maryland lottery pick 3 results over the three day period were 956, 141, 555, 280, 885 and 471 respectively.

You will notice the winning pick 3 number 555 in Figure 1A. At that point, you are still observing what is going to happen. It did not offer you an enough clue to risk your money. The pro money is hard earned and must be wagered based on a good risk reward ratio.

If the risk of losing is too high do not wager on that particular game.

The MD lottery continues dropping through the fourth game when they played 471. If you count Figure 1A down through four games you will get 471. This confirms that the trend is in place.

The real question now becomes, are they going to continue and play 829 or go back to 623 and 393. Those three are the numbers in front of the line. The next thing will be to start eliminating the weak ones.

The trend is clearly showing 829 to be the first in line followed by 623 and 393. The big question with 829 would be, are they are going to play it the way it is since 471 played straight? The second question is, could they have played it as 280 that dropped right after the winner 555? If it played after 555 as 280, which would have been the position of 393 in Figure 1A, does that mean that they are going to play 393? The next question would be, could they play the 829 as potential repeat of 280 or drop it as 829. What are your options and how do you maximize your winnings in the seeming difficult situation?

Those are the questions the pro player will ask before wagering on any lottery game.

The first thing you need to do is check Figure 1B and 1C to see if the trend transferred to either of the other two thereby saving you the trouble of searching. In this case, it did not. Please note that Figures 1A and 1C are two different groups. They are, however, related to Figure 1A to the point that you can use one group as a mirror to the next one.

The three groups could be playing simultaneously in three different states. You can use one state trend for any place you are wagering on. It does not matter if the three groups played at different times.

If Maryland played 623 and 471 for midday and evening yesterday followed by 829 the next midday game, it clearly shows you the direction of the trend. If Florida dropped 296 the same day Maryland played 829, your best bet for Florida will be 420. You can see that Florida continued the trend with Figure 1B.

That does not mean that Florida might not go to the other direction with 715. The trend can go in any direction. The group in any trend is a complete circle. In this case, if you are spending just $1 consider 420 based on the existing trend before 715.

The reality is that the house would not expect you to expend this much time looking for the winner or have this book, as a matter of fact.

We are yet to completely answer the question of playing 829 or the number to fill the gap. The first thing you should know is that those three pick 3 numbers 393, 623 and 829 are going to play. If the 280 played in place of 829 it means it could take up to one month for the 829 to HIT. If you add one and nine without carrying over it would give you zero (0) thereby creating the 280. The one added is an indication that it could take up to one month before 829 drops.

In this scenario, the options will be to start playing the 829 if you choose with the mind that it could take upwards of one month before it drops. If you choose to go to that route consider playing it with 393 and 623. Those three are exhibiting the same behavior. It means that they are going to drop over same period of time.

As you know, those three enveloped the 555 and 471 that played from the same group. They are in the same trend and are bound to drop.

If you choose that approach you must calculate how much it is going to cost you over that period of time. You could mitigate some of that cost by cashing out on any of the three that plays before that time. That of course will make sense if the money won at that point is far greater than the cost.

Another option would be to play parlay on 829. You could play parlay on 829. The parlay of course is meant to give you some house money for much bigger payout later. The parlay in this case will produce twelve numbers. The primary reason is because 829 is six ways since it is single numbers. You will need to cover the front and back pair. It means that if two of the three numbers show up you will win a ten percent of what the traditional win would have been.

Do not put a lot of money in this instance. The trend is right on 829. If you choose to play it, your first pot of call will be to play 829 straight (exact) and box (any order) and cover the rest of the play with parlay.

The parlay (front and back pair) in this case will look like this,

X82, 82X, X28, 28X, X29, 29X, X92, 92X, X89, 89X, X98 and 98X.

In this case, the parlay produced twelve sets. I would advocate playing it with 50c, which would cost you a total of $6 on the parlay plus $1 spent on the straight and box for a grand total of seven dollars.

The question at this point would be, is the potential payout justified in spending seven dollars? The answer would be yes based on the current trend of figure 1A.

You do not, however, need to spend too much money based on the underlying questions. The potential payout on the parlay will be $25 if any of the numbers HIT. The potential reward will be three to one. That will make it a good game to play.

Lo and behold, they played 392. This would have made $25 for you minus the $7 cost. Your initial $7 has turned to $18 thereby giving you some house money to play with. This gives you room to put more money in the next game when the trend is right.

If it takes you another day or two to identify the next winner stay with it.

What sets this book apart from others is that you can check the same trend in any state or country of your choice and still come out well. Trends do not lie and it is your friend.

Could there have been a better way to play that trend and make more money? The answer is unequivocally yes. You are the pro and your job is to make money.

The scenario above is not the first choice of the lottery pro player. It is good for those that do not want to look through the book and capture the trend. The pro player must be willing to put the simple effort of looking through the book and identifying the trend.

The pro player does not wager if the potential reward does not far outweigh the risk.

Are you going to leave the chance of winning with 829? What if 829 drops today?

How are you going to make sure that you win some money if you wager on this game?

If the house plays something completely different how do you catch it?

If you miss 829 are your watch list set to win the next one?

You do not play to lose. The pro players play to consistently win. That is what sets you apart.

The answers lie in the next chapter.

Chapter 6

Lottery Wagering Know How

Let us explore the trends further. This exercise prepares you more for pick 4 that will come in later chapters.

Figure 1

D	E	F
956	271	680
139	789	141
916	134	265
608	280	793
919	829	161

522	770	280
708	624	891
514	112	722

You can go back as far as possible in the case of Maryland or any state of your choice and look at the past results against the above trends.

Let us explore the trends further with Figure 1D, 1E and 1F through a different state. This is meant to cover different angles of the lottery game to give you enough tools to tackle whatever trends you are wagering on.

Let us look through Georgia lottery results.

A quick look at Figure 1F will show you that Georgia lottery is playing that trend over the last three weeks. The three weeks are based on the time of writing this page, which is 6/7/2011.

Georgia lottery started that trend with 793 on 5/13/2011 and continued three days later with 161 on 5/16/2011 and 082 on 5/21/2011. You can see how close they are dropping right after each other.

A good observer will catch some of the winners.

You can actually win a lot of those when you have other trends to watch as well. The other trends might confirm the ones you otherwise may not notice from Figure 1F.

The trend continued with 918 on 5/24/2011 that repeated straight as 891 on 6/2/2011. The next Georgia Cash 3 that followed was 272 on 6/6/2011. You can see that the 272 did not HIT straight. It, in essence, is playing the same role that 918 played.

Georgia is still going to play 722 to complement 891. The 722 is being pulled down by 891. The trend will continue through the rest of the group.

If you look from Figures 1A through 1F you will notice that Georgia played nearly every number in the group within the last one month. Those are opportunities for the pro player. A keen observer will make good money here.

The same thing is applicable to every state or country.

Figure 1

A	B	C	D	E	F
122	229	212	956	271	680
555	557	555	139	789	141
393	932	937	916	134	265
623	235	267	608	280	793
471	715	743	919	829	161
829	296	283	522	770	280
442	420	448	708	624	891
962	627	964	514	112	722
527	277	257	XXX	XXX	XXX

This information will prepare you for the next time Georgia or any other state is playing that trend. Use the above template and others to win every pick 3 game. We will now touch on Figure 2 briefly. The initial exercise is to study the above table on your state lottery results and one other state. You should look through a two month period. Get ready to elevate yourself.

Chapter 7

Mastery Of The Trend Is Winning

You need to master how to look at several trends at any given time. Those trends could give you clue based on what your state is playing or that of other states. There are times the games will follow each other and times you may notice some gaps. The gaps from one group could be made up by trends from another.

I will put down several trends and show you how to use it in any lottery game of your choice.

Figure 2.

A	B	C	D	E	F	G	H
494	106	207	174	850	425	218	120
381	632	276	485	780	102	327	473
865	649	759	847	506	957	798	119
499	873	015	515	935	419	569	294
204	602	899	553	111	009	260	321
142	688	298	715	680	892	348	897
692	522	533	491	919	153	495	329
852	179	090	367	528	692	212	642

From Figure 2D, you will notice that Maryland lottery results dropped 471 on 6/6/2011 followed by 485 on 6/7/2011. The Maryland lottery winning pick 3 numbers are trending right after each other. The next day, 6/8/2011 Maryland lottery played 419. The 419 clearly deviated from Figure 2D. What happened here is that it created a new opportunity for the readers who follow this book clearly.

The trend seems at this point to have continued on Figure 2F with 419. The existing trend was supposed to have continued with 847. They jumped it and played 419. They came back the next day on 4/9/2011 and dropped 838 in place of 847.

Please pay close attention here. The 838 played in the position of 847 as a decoy. The two numbers are the same here towards the completion of the existing trend. If you take one from eight and add it to the three in the middle the result will be 847.

```
    8     4     7
  _       1  +  1
  _____
    8     3     8
  _____
```

You will notice that the sum in each case will be 19

8+4+7 = 19

8+3+8 = 19

What this means in the lotto pro world is that this is an opportunity to make some money. The two scenarios here are that the winning number 838 produced a potential opportunity anywhere in the book that has 838.

The other opportunity is that you should go after the known trend which in the case will be 515. The pick 3 number 515 is the game below 847. You are wagering on it because the trend continued as demonstrated above with the sum of 19.

The professional approach will be to play the pick 3 number as well as parlay. The parlay will give you six positions since it is a double number. The result of the parlay will look like this,

X55

55X

X51

51X

X15

15X

The parlay will pay you some money if two of those numbers show up. Some states do call it the front and back pairs.

Lo and behold the very next Maryland lottery result came as 513. The 51X would have won in this case thereby giving you 417% return on your money. That is just one return in a typical day. That is why it pays to carefully observe the trends as laid out in this book.

That opportunity is for the pro to grab. You will find endless trends if you put the effort. In the above case, every $6 would pay you $25. Some people may think that it is small money. You have more opportunities now that the house play more than once every day in most places.

They do to make more money. As you may well know, the house does not lose except from the pro that knows what they are doing. They may avoid the crowd, but they will fall into you.

You will remember that Maryland lottery played 419 on 6/8/2011 from Figure 2F. That particular column is trending upwards. Another look at Figure 2F will show you that Maryland lottery played 254 on 6/11/2011. That will be exactly four days from the day they dropped 419. The next pick 3 number after the 425 is 692. Maryland lottery played 693. This will make another parlay winner thereby putting more money in your pocket.

The 692 is still going to play. You can see that it is between 425 and 153. Those two played recently and the 692 is going to HIT. You are able to make three times the money wagered on the parlay 69X thereby giving you the house money to work with.

I once played X73 with $10 and the cashier at the retail outlet thought that I was crazy. She changed her position the next day I went to her counter to cash $500. There is no limit to how much you can make. This book gives you opportunities where none exists.

You could have winning tickets every single day if you follow the instructions in this book.. That is what makes you a pro.

That column presented an opportunity once more for the pro player. You cannot leave home without this book. You could be leaving money on the table.

You will consistently play with confidence whenever this book is in your hand. Every time a new pick 3 number plays brings another opportunity elsewhere in this book. You can only see it when you take a careful look.

That is what lottery smart money does. The potential payout will justify the effort.

Figure 3

A	B	C	D	F	G	H	I
612	381	602	717	418	182	767	663
831	498	242	956	379	532	725	256
286	653	599	828	135	630	777	535
384	625	486	265	527	329	309	980
012	908	207	472	467	060	250	949

057 714 942 826 706 035 229 078

249 908 240 041 812 796 667 277

296 342 101 870 128 640 706 104

653 847 696 664 553 461 216 267

You can follow in ascending or descending order based on the direction of the state you are watching. For instance, you will notice that Maryland lottery played 471 followed by 485 and 419 in the first week of June 2011.

You will find the first two winners in Figure 2D followed by 419 in Figure 2F. In this case, the winners are trending downwards. The existing trend shifted to Figure 2F making the 957 and 009 potential winners.

The same trend could have continued in Figure 1, 2, 3 or any other one. All you need to do is count based on the existing trend to where the current winner is.

You can count with your home state or any other state or country that is trending in the direction.

This knowledge gives you endless winning potential. If you do a careful study of the trends, the winners will be right in front of you.

Please do take advantage of parlay. It gives you the house money to wager with. Do not wager on any game if the reward ratio does not far exceed the risk of loss.

The pro player does not wager for fun. This is business.

If you meticulously follow the trend, you will find the games among the tools in front of you.

This tool decimates the odd calculations because the trend could continue in any of the groups, and you will still be able to catch it.

Figure 4

A	B	C	D	E	F	G	H
741	286	053	756	113	197	701	444
149	733	207	535	575	295	335	291
908	001	670	200	630	571	220	043
555	798	098	638	309	550	601	929
000	537	519	156	999	704	353	547
008	703	337	995	482	558	115	219
709	372	327	043	446	211	812	187

There are times when a certain opportunity will warrant carrying some numbers longer. An instance of that will be some pick 3 numbers that played in New jersey recently that was trending from Figure 4G.

Let me bring down figure 4G to buttress the position. You can do the same with any of the groups that you are following.

Figure 4G

701

335

220

601

353

115

812

New Jersey lottery played 353 on 5/2/2011. At that point, it would not be enough for you to do anything, unless you justify the next winner based on trends from another state or group.

They followed up with 202 on one month anniversary of 353. That date would be 6/2/2011. That is enough to show the pro player that they need to pay attention to Figure 4G.

The very next day 6/3/2011 they dropped 601. You will find all three winners in the Figure 4G group. That group shows you that 353 is a repeat candidate.

The winner based on that trend is 353. It is bound to repeat. There is a good chance that it will come back as 353 based on the fact that 220 played as 202 and the other justification is that the initial two played straight.

You do not need to wait until the next anniversary to play it. You could start right now. If you catch it in any order it will pay $160 on $1 box. If you HIT the pick 3 straight as well, you stand to collect about $660.

The potential reward for New Jersey bettors is high. That number is yet to play as at the time of writing this page, 6/9/2011.

This same group could play in any other state at some point. The only people that will understand the group are those with this book. One pick 3 number worthy of note is 808. The pick 3 number 808 is one to watch for the 353. It played prior to the initial 353 and did as well before 202 and 601 dropped. It may come before the next 353. Those in New jersey should consider playing it, even if 353 shows up first. You should look through the groups in this book before playing.

You have seen different ways of playing the pick 3. I will revisit other pick 3 methods later in the book. This book will not leave any stone unturned. Let us look at some pick 4 games.

Chapter 8

The Pro T Set System

This is the bridge between pick 3 and pick 4. The Pro T Set System plus Cornucopia Methods means that you become unstoppable. There is no lottery game you cannot win on this planet. Do not lend this book out because they will never return it to you.

X07	7946	225	2830
682	6423	0X0	3402
933	8462	855	2548
375	1941	032	3131
179	1584	164	4639

917	4439	589	7595
589	7591	849	1744
311	6446	194	1179
254	8032	138	4623
303	4028	758	2642
552	2528	393	0077

The pick 3 might not move in tandem with the pick 4 at all times. You can use each group to identify the potential winners. You can transpose with the Cornucopia Methods as needed.

The figures in this book cover every possible trend that played and ones yet to come.

You will be smiling all the way to your wallet if you follow the instructions in this book.

The methods are already in place.

The system has already been tested.

There is no angle in any lottery result that is not covered in this book.

The tool you will need to bring is patience and mastery of this book.

You must study the book relentlessly.

Do not forget that the pros do constant practice.

This book will put money in your pocket if you follow the instructions.

Chapter 9

The Pro Source Of Major Money

This book will show you different angles to winning pick 4 games. The book will shatter the pick 4 myths. It will give you the tools you will need to challenge the odd of 1:10000.

The pick 4 runs in four, six, twelve and twenty four way positions.

Let us take a quick look at one pick 4 game that have a four way position. The four way means that it can only be rotated in four different ways, e.g. 6000 can only come in four different ways.

Single pick 4 numbers like 6789 will come in twenty four different positions.

Take a look at 6000 in different positions and the resultant changes in each position.

Group

A	B	C	D
2742	7878	0315	2855
0006	0060	0600	6000
3393	4507	3226	0717

Another close look on the above groups will show you that each group is completely different although 6000 is common to the groups. The calculations will come out different each time. If either the top or last pick 4 numbers in any of the groups' changes by even adding just one, the entire equation will change.

I am sure by now you may be wondering how these complex calculations are made. The real concern should be how to get around the pick 4 odd calculation of 1:10000.

The above Group sample is part of what will separate you from the crowd. You cannot find it in any of the other so called lottery books. They simply do not know how and cannot do the calculations. It cannot be done by software either.

The software is not capable of doing this calculation.

The reason you will know that you are at the right place is that these numbers will be tested against actual state lottery results. You should feel free to check them through any state or country.

You will have tons of tools to win any pick 4 games. You will be able to work out winners by yourselves.

Chapter 10

Pick 4 Trend

The most money in pick 4 is made when others are not looking. It is made when they cannot recognize the opportunities. It is made with consistency. The consistency comes with knowing what to look for. The odds of winning pick 4 is 1:10000.

You have to master how to get around the odds to come out ahead. You have to know the weaknesses of the pick 4 games, recognize it when the opportunity presents itself and take advantage of it. In the world of investing, so many people will jump through the roof if they can get annual returns of about ten percent.

Most of them do not even come close. The pick 4 lottery game, on the other hand, gives you the opportunity to make potential five thousand percent return. Think about that for a minute. That percentage is so huge that it is worth mastering every tool in this book.

Every pick 4 winning number produces another one. Percentage returns compounded is the key. If you can capture a good percentage of that 5000% return, you will be in good shape. Your job is to consistently make good return on any money you wager in lottery games. Do not forget that it is a business and not just fun.

The average lottery player out there have the possibility of losing majestically placed in their mind before wagering. They will be happy if the luck ran into them occasionally. That position by itself alone is a loss.

The hard fact here is that every game has a weakness. You are here to master those and exploit them. You have before you the best lottery book ever written.

In any discipline there will only be one best. This book puts you in that position.

You should feel free to ask any lottery book author to challenge this book and you against any book in the market. This book will put you in the elite class.

Make sure you read every page in this book.

The pick 4 odd calculation of one in ten thousand can only be justified to the extent the house plays any group in a particular trend over a stretched period of time. That span makes it difficult for people to understand thereby settling for the odds.

You are bound to lose when you are within the confines of the odd calculations.

You will never look at the odds' calculations the same way after reading this book.

Pick 4 Trend Group 1A

Let us take a look at a group of pick 4 numbers in one trend through DC lottery.

5203   DC lottery played it on 5/14/2011 and dropped 1827 on 5/15/2011

4458   DC lottery played it on 5/23/2011

6809   DC lottery played it on 6/9/2011 and dropped 1278 on 6/10/2011

0696

4745

6901

0285

8547

6987

The above trend will show you that DC lottery played the first three as at the time of writing this page, 6/11/2011. You will notice that all those three did not play immediately after each other.

The trend started with 5203 on 5/14/2011 followed by 4458 on 5/23/2011 and 6809 on 6/9/2011. You will notice that the first two played a little over one week apart and the third one 6809 dropped almost three weeks later.

The people without this book will never understand the relationship between all three and the rest of the group. They are all in that one group playing in the trend.

If you look through DC lottery results, you will notice that 1827 played in just one day after 5203. That date would be 5/15/2011. That same pick 4 numbers came back as 1278 on 6/10/2011. That again will be one day after the 6809.

I will discuss the pick 4 parlay later.

The initial three winners for DC lottery shows that they are playing the group downwards. They could play any of those. Your job is not to catch all of them with precision. Your job is to make good money with the group or any other group in a particular trend.

The above group could be playing in your state by the time you are reading this book, or they could serve as a mirror.

If you start playing the remaining six at $1 straight every day, it will cost you about $2300. The initial three that played dropped two of them straight.

If you win only two of those by the end of the year that will give you $10000. That is a very good return on $2300 wagered. The beauty here is that this book made those numbers available. This method works best for those readers who prefer to play the same set of pick 4 numbers every day.

If you spend a total of $23000 over one year and win at that rate, you will be come out with $100000.

If you are getting very high returns it is only logical that your money will multiply based on what you put in. You can have a good number of those groups and wager within your budget.

Pick 4 Trend Group

B	C	D	E	F
1101	9782	2202	6957	8305
6532	3601	9961	1884	0550
5225	6500	1882	3171	2174
3037	2207	5741	0360	6721
7646	5558	8665	1510	2981
4152	0552	6207	6564	9132
1680	9947	0395	5208	0126
4756	9066	8686	9200	8546

6925 **5706** **2206** **4216** **3793**

Chapter 11

Monthly Pick 4 Templates

Please do a careful study of Monthly Pick 4 Templates and see explanations below with explicit details on how to use it. You will need this tool and more to win pick 4 consistently. The pick 4 run in circles, and the winners will always be at your fingertips.

	A	B	C
1)	9785	1012	5003
2)	1202	8245	4240
3)	6570	4032	7185
4)	6205	2518	3012

5) 1095 4132 7060

6) 2449 1711 1818

7) 0214 4086 2257

8) 8159 6150 7994

9) 4889 2369 7069

10) 3051 6505 3198

11) 4928 8459 9690

12) 2646 2609 4013

13) 8017 2535 4457

14) 3448   6174   2225

15) 2274   6631   4471

16) 0616   2124   5823

17) 1820   1262   7719

18) 8106   6002   7401

19) 2200   6742   9287

20) 3556   9263   0439

21) 6168   2752   6900

22) 6716   5582   7498

23) 1513   6361   0236

24) 2820   7199   4627

25) 7162   1632   9028

26) 2221   5189   1241

27) 2050   2520   8955

28) 7550   0007   8724

29) 9263   5618   0644

30) 5919   1759   2954

31) 6608   6168   0879

Take another careful look at Monthly Pick 4 Templates. You will see that column A number 10 had pick 4 number 3051. A look at that column through the eye of Florida lottery will show you that Florida played the same pick 4 number on 3/2/2011. On that same day Florida lottery dropped 0588.

The monthly Pick 4 templates has 4889, 3051 and 4928 on the column A numbers 9, 10 and 11 respectively. The professional question here should be, what is the correlation between 0588 and the 3051 that played the same day?

We have 4889 in column A number 9. If you reposition that pick 4 as 4988 and add one (1) to the 9 it will result in 50 thereby creating that winner. It shows you that the trend is in place based on the Monthly pick 4 Templates.

This is an opportunity that the pro player cannot miss.

This does not in any way disregard the 4889 as part of that group. I will buttress that position by telling you that 9488 actually played a month earlier on 2/13/2011 thereby clearly making the 4928 the next in line to drop.

The demonstration shows you that 4928 is yours to win.

There are two ways you can win. I will discuss one now and introduce the other part later in the book.

The third pick 4 numbers HIT three days later as 9482 on 3/5/2011. The box winning alone on that would give you over 1650% return on your money. It is worth mastering this book with that type of return.

This book is meant to identify similar opportunities every single day. You have enough tools to capture winners every day. The money is in front of you. They cannot avoid you. This is a battle against conventional odds.

Let us take another look at Monthly Pick 4 templates through Massachusetts lottery. Please look at section A numbers 18, 19 and 20. You will find 8106, 2200 and 3556 in that order.

The Massachusetts lottery calls their pick 4 lottery, The Game. We are going to use the popular name here which is pick 4 to accommodate every reader.

Those pick 4 numbers lined up in this unique system as follows,

6553

8106

2200

Massachusetts lottery started the trend with the pick 4 number 8106 on 3/26/2010. They capped that month with 6553 on 3/31/2011. What is left to complete the trio will be the pick 4 numbers in the middle. **That GEM is 2200**. It is obvious that it is going to HIT based on the fact that it is in the middle of the trend.

The Massachusetts lottery played it on 5/5/2011. That will be mere five weeks from the time the last one played. You cannot miss this opportunity as a pro. The potential payout is huge. The crowd wagers on lottery with very high urgency. They are preoccupied with mighty hope of being lucky to the point that most does not analyze the game. They do not look at the risk reward ratio before playing any game.

This game won in exact order pays $5058. That is more than five thousand percent on $1 wager. I do not see any reason you should miss that opportunity when it presents itself.

The pick 4 templates clearly showed you that it is coming based on the trend.

The three winners played in less than two months. How do you fit the ten thousand odd calculations in that equation? You cannot. It is better you spend the time combing out the winners in the book.

You might be tempted to ask if this opportunity is available every day. You should, in fact, ask. Let me do it for you. I will give you the answer as well. You cannot ask for more than that. You should run with it as a pro. Here comes the question and answer.

Question: **Is it possible to win pick 4 everyday?**

Answer: **Unequivocal Yes.**

Chapter 12

Cornucopia Methods

The answer to winning every pick 4 games.

All pick 4 numbers must travel at certain trend. If and when the trend changes, the calculations must be redone to capture the new trend. You can win pick 4 games where others may not see based on the Cornucopia Methods and the existing trend.

This is possible because the prior results serve as a mirror to the next winners thereby creating relentless opportunities. The parlay method on pick 4 games make the opportunities endless.

There is about ten thousand opportunities in pick 4 games. You have the opportunity to make money with one or two at the most in your home state daily, unless of course you have your own lottery club that covers every state and country.

You should entertain the idea.

The Cornucopia Methods is designed to cover any gaps that might exist. This means creating the winners if you do not find it in the Pick 4 Templates.

You will use the Cornucopia Columns to create the winners based on the existing trends. I will create and give you examples as well as parlay. Please do the same exercise with three different states. This is business and you will need the skills to make steady money.

Buckle your seatbelt and take a ride to serious money.

Cornucopia Methods (CM)

A	B	C	D	E
26	76	74	36	75
56	02	60	26	56
23	26	21	46	91
53	70	32	01	10
06	11	18	84	26
73	11	41	47	60
35	73	25	31	34
67	42	07	29	55

Cornucopia Methods is the trends that are already calculated to follow each other thereby giving you the advantage to transpose any group and recreating new trend. The essence is to recapture trends that otherwise would not be there. You can do this to virtually any pick 4 games thereby eliminating the need for ten thousand odd calculations.

This in turn enables you to play parlay on any pick 4 game and create opportunities that would not be there or follow the actual trend. The result is endless winnings.

Let us take a look at Cornucopia methods in play.

From the eye of North Carolina Lottery.

North Carolina lottery played 8000 on 4/27/2011. This winning pick 4 numbers are not in any of the templates in this book. You are not necessarily going to walk away from it simply because you don't see the pick 4 number 8000.

Your job as a pro is to look for and create opportunities. You have looked through the templates and did not see 8000 at this point. The fact of the matter is that you cannot possibly fit ten thousand pick 4 numbers in this book or any other lotto book, for that matter. Frankly speaking, there is no need for it.

The reason there is no reason is because the Cornucopia Methods is in place to address that minor problem.

Please take a very close look at the Cornucopia Methods Colum A through J and continue reading afterwards.

F	G	H	I	J
29	62	87	95	08
87	67	65	95	66
80	38	76	60	07
53	67	13	40	11
00	85	17	26	61
62	21	56	11	55
76	42	42	32	74
70	XX	XX	XX	XX

The next job is to recreate North Carolina winning number 8000. This winner of course could be from any state. It does not matter where the number plays because you have the CM in your hand. Now in your quest to recreate 8000, the first place to go to will be CM Column A.

You are going to start from CM Column A because it has the numbers 80. The next place will be two steps below that because you will find the numbers 00 thereby creating 8000. The primary question will be, how would this help you win the pick 4 games in this case?

Let us take a look at the result after you have transposed the numbers.

The New (CM) Created Trend

A	B
2980	2976
8753	8762
8000	8000
5362	5353
0076	0080

The North Carolina lottery results played as The New (CM) Created Trend Column B. The difference between the A and B groups are that the last pair was turned upside down. I reversed the position to properly identify the direction of the trend.

Please do not forget to do the same thing in your exercise or when you are looking for actual opportunities. This is business and should be done as such.

North Carolina lottery played 8000 on 4/26/2011 and followed the next day with 5353 on 4/27/2011. You can clearly see both winners in The Newly Created Trend column B. This opportunity could not be possible without the Cornucopia Methods.

The larger question would be, how could this benefit you?

At this point, the 8000 and 5353 is given. The Column B above shows 8000 as a solid repeat candidate. The trend is going upwards and the next winning pick 4 numbers will be 8762.

In your quest to ascertain playing the 8762 you will find out that it actually played as 6287 on 2/26/2011. You would have worked out 5353 if you have that information that might have made some money for you.

You could still make money with parlay. The parlay is where you keep two of the pick 4 numbers constant and replace the other two with the CM pair. Once they are replaced, you will then play each of those with numbers zero (0) through nine (9).

Now we found out that 6287 already played on 2/26/2011. In this exercise, it became clear that 2976 equally played. The 2976 came as 6972 on 3/17/2011 and repeated as 6792 on 3/28/2011.

The New (CM) Created Trend Colum B shows 8000 coming back as 0080. This means that this group is playing as repeats, a position that is now buttressed by 2976. This creates a lot more opportunities considering the fact that most of the pick 4 numbers in this group pay much higher.

The next number to HIT based on the trend will be 8762. It is bound to come back to compliment the rest of the group.

The 0080 as potential repeat could drop as 0400 to complete the one that played on 4/30/2011. If you add the two together it will sum up to 0800.

The Cornucopia groups could be transposed to any of the pick 4 games thereby creating more opportunities. You can use it to play parlay. An example of parlay will be if you have the pick 4 number 1234 and the Cornucopia Number 79, you will keep two of the original pick 4 numbers constant like 12 or 34 as the case may be. Let us assume that the number you are working with is 1279, the parlay will now be 127X and 129X. The X represents zero through nine.

You may play that number following the last pick 4 winning number that resulted in the creation of the new numbers. If the parlay does not play consider playing them from four days after the initial result. They tend to drop not long after that.

Please follow the instructions carefully if you fancy playing pick 4. I have tested these games over the years. You do not need to reinvent the wheel.

Let us take one more look at the power of Cornucopia Methods. Please go to CM Column F through the lenses of Maryland lottery pick 4 results of 3/28/2011. The Maryland lottery dropped 9462 on the day in question.

In recreating the trend to explore opportunities, you will follow the instructions as described in previous chapters to arrive at the new winning pick 4 number 9462.

The CM Colum F will recreate that pick 4 number thereby forming the trend below,

3007

5167

9462

5900

4853

You will obviously notice that 5167 played the next day 3/29/2011. This shows you that the group is trending upwards. You are not going to jump in and start playing 3007 without checking prior results.

The house could come back and play the 5900 below that. Your concern should be coming up with the winning numbers.

In checking past results, you will find out that the house actually started the trend with 5900. It HIT initially as 5090 on 1/13/2011 and came back as 0059 on 2/11/2011. The 5900 dropped straight on 1/25/2011.

The larger picture here is that the 5900 lead the way with three HITS. This means that the rest of the group is going to drop three times. Please remember that groups tend to exhibit the same behavior.

The next pick 4 winners that followed were 9462 on 3/28/2011 and 5167 on 3/29/2011. You can see that the three winners actually played straight. That is an opportunity that should not be taken lightly for the fact that it gives you upwards of 5000%.

That is huge to have produced the three in such a short period of time. Most trends will start with one of the groups and pause for upwards of two months before resuming. In cases, they dropped two in a relatively short period of time be prepared to carrying the third one for up to two months.

You have to be patient when wagering on pick 4. If you identified the three that won so far, that would pay you $15000 in one year. Your total cost would be about $360.

We have not exhausted the trend. The next pick 4 numbers in the group are 3007. You will find that it dropped as 7300 on 4/7/2011 and repeated as 0730 on 4/25/2011. It is going to come back for the third time since the lead pick 4 5900 played three times.

I will let you find if 4853 have played through the instructions given in this book. You should create trends and check the results. Identify the opportunities based on the trends.

Your opportunity to win pick 4 games are endless.

Do your homework and start winning. You are the pro. Go win some money.

The Cornucopia numbers are mathematically worked out to follow all lottery trends.

Cornucopia Methods has never failed and will never fail.

This book carefully mastered puts you above the realm.

All rights reserved. No part of this material shall be reproduced or transmitted in any form by any means, including photocopying, recording without written permission from the publisher. The numbers in this book has been carefully worked out with the rigors of Lottery Pro Player and our readers in mind; however, these are recommendations only.

Made in United States
North Haven, CT
31 March 2024